WILDLIFE OF MEXICO

SOUTH OF THE BORDER

Mel Higginson

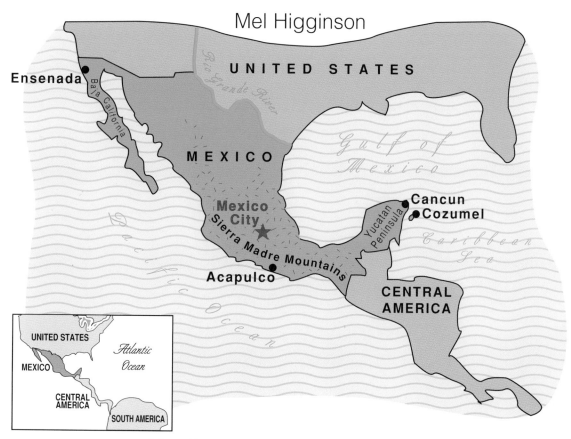

The Rourke Book Company, Inc.
Vero Beach, Florida 32964

Edited by Sandra A. Robinson

PHOTO CREDITS
© Lynn Stone: cover, pages 8, 10, 12, 15, 17, 18, 21;
© Steve Bentsen: title page, page 4; © Gary Vestal: page 7;
© Aldo Brando: page 13

Library of Congress Cataloging-in-Publication Data

Higginson, Mel, 1942-
   Wildlife of Mexico / by Mel Higginson.
   p. cm. — (South of the border)
   Includes index.
   ISBN 1-55916-055-1
   1. Zoology—Mexico—Juvenile literature. [1. Zoology—Mexico.]
I. Title. II. Series.
QL225.H54 1994
591.972—dc20
                                    94-14999
                                    CIP

Printed in the USA
                                    AC

TABLE OF CONTENTS

WILDLIFE OF MEXICO

Mexico's wild animals live in deserts, jungles, brush country and mountain woodlands. They live in swamps and rivers, in the sea, and along the seashore of Mexico's coast.

Mexico is in the southern part of North America, between the United States and Central America. Many of the same kinds of wild animals that live in the United States live in Mexico. However, Mexico has many kinds of animals that also live in Central and South America, such as the **tapir** and king vulture.

The peccary, or javelina, is found on both sides of the U.S.-Mexico border

INSECTS

Mexico has one of the most amazing insect collections in North America. That is because 200 million monarch butterflies gather in Mexico's Sierra Madre Mountains each winter.

The monarchs reach the Sierra Madre Mountains after flying hundreds of miles south from the United States. The monarchs cover branches, tree trunks and leaves in orange, black and white.

After spending the winter in the high, cool Mexican forests, the butterflies **migrate,** or travel, northward.

Wintering monarch butterflies hang like leaves from the branches of trees in the Sierra Madre Mountains

REPTILES

Mexico's **reptiles** live in wet country and dry, high country and low. Snakes and lizards are among the most common reptiles. Two of Mexico's most interesting snakes are rattlesnakes and boa constrictors.

Mexico is the only country in the world with two kinds of poisonous lizards. The toothy Gila monster and the beaded lizard, a close cousin, live in deserts.

Endangered American crocodiles live in Mexico, along with several kinds of sea turtles. Sea turtles are also in danger of becoming extinct, or disappearing forever.

The Gila monster has a poisonous bite, but poses little danger to people

BIRDS OF PREY

Mexico has a wonderful variety of birds of prey — hawks, eagles, vultures and owls. The osprey, or fish hawk, hunts along the coasts. Golden eagles hunt rabbits and other small animals in open country.

Mexico's most colorful bird of prey is the king vulture (see cover). These big, handsome birds don't kill animals as other birds of prey do. Vultures scavenge — they eat dead animals they find.

*The caracara, a bird of prey, is
also known as the Mexican eagle*

The jaguar is the largest of Mexico's predators on land

The black and white hawk-eagle lives along the wooded riverbanks of southern Mexico

LAND BIRDS

About 600 kinds of birds live in the United States and Canada combined. Over 1,000 kinds live in Mexico. Most of Mexico's birds are "land" birds. They live and hunt in dry areas.

Mexico has many of the same kinds of birds as the United States. It also has different birds, such as **toucans** and 19 kinds of parrots. Some Mexican birds have really strange names, like flower-piercer, bananaquit, motmot, woodcreeper and grassquit.

Scarlet macaws live in the rain forests of southern Mexico

WATER BIRDS

Mexico's swamps, lakes, rivers and more than 6,000 miles of seashore are habitats, or homes, for water birds. Water birds often eat plants or animals that live in water.

Mexico's largest water birds are two kinds of pelicans — brown and white. Some of the other large water birds, like herons and egrets, have long legs for wading. These birds use their long, sharp bills to spear fish.

Many other kinds of water birds, such as gulls, terns, ducks and geese also live in Mexico.

Mexico's largest water birds, white pelicans, spend winter along Mexican coasts

SEA MAMMALS

All sea mammals spend part, or most, of their lives in the ocean. The porpoises and whales spend their entire lives at sea. Mexico's fur seals and huge elephant seals crawl ashore on sea islands to rest and give birth.

The Sea of Cortes on Mexico's west coast is the winter home of California gray whales. The 45-foot whales give birth to their calves in warm, sheltered sea **lagoons.**

Elephant seals haul themselves ashore on Mexico's sandy sea islands

PLANT-EATING MAMMALS

Plant-eating mammals can be small, like mice and rabbits. They can be large, like deer and bighorn sheep. Several kinds of plant-eating mammals, small and large, live in Mexico. Some of the largest — deer, pronghorns, **peccaries** and desert bighorns — also live in the southwestern United States.

Mexican plant-eating mammals such as spider monkeys and howler monkeys, tapirs, and **agoutis** do not live north of Mexico.

Plant-eaters are **prey,** or food, for **predators,** or meat-eating animals.

Desert bighorns live in the dry country of northern Mexico

MEAT-EATING MAMMALS

Six of Mexico's predators are kinds of wild cats. The largest is the jaguar. A jaguar can weigh over 300 pounds. Mountain lions, bobcats, jaguarundis, ocelots and margays — small spotted cats — also live in Mexico.

Mexico is the home of several other furry predators, including foxes, coyotes, black bears, ringtails, badgers, skunks and others.

Mexico's bats, armadillos and anteaters eat insects.

Glossary

agouti (ah GAH tee) — rabbit-sized rodent of Mexico and Central and South America

endangered (en DANE jerd) — in danger of no longer existing; very rare

lagoon (luh GOON) — a long channel of water attached to a larger body of water

migrate (MY grate) — to travel from one place to another at the same time each year

peccary (PEHK uh ree) — a small, wild pig, also known as a javelina

predator (PRED uh tor) — an animal that kills other animals for food

prey (PRAY) — an animal that is hunted for food by another animal

reptile (REHP tile) — a group of cold-blooded animals including lizards, snakes, turtles, alligators and crocodiles

tapir (TAY per) — a large, short-legged animal with hoofs and a long, down-curved upper lip

toucan (TOO kan) — a fruit-eating bird with a large, long beak

INDEX